Why Science Matters

Using Genetic Technology

Andrew Solway

Heinemann Library
Chicago, Illinois

Editorial by Andrew Farrow, Megan Cotugno, and Harriet Milles
Design by Steven Mead and Q2A Creative Solutions
Original illustrations © Pearson Education Limited
Illustrations by Gordon Hurden
Picture research by Ruth Blair
Production by Alison Parsons

Originated by Heinemann Library
Printed and bound in the United States of America, North Mankato, MN.

14 13 12
10 9 8 7 6 5 4 3

Library of Congress Cataloging-in-Publication Data
Solway, Andrew.
  Using genetic technology / Andrew Solway.
    p. cm. -- (Why science matters)
  Includes bibliographical references and index.
  ISBN 978-1-4329-1837-8 (hc) -- ISBN 978-1-4329-1850-7 (pb)
  1. Genetic engineering--Popular works. I. Title.
  QH442.S653 2008
  660.6'5--dc22
                    2008017749
102012
006969

Acknowledgments
The publisher would like to thank the following for permission to reproduce photographs: ©Alamy **p. 31** (Sean O'Neill); ©Corbis **pp. 5** (Holger Winkler/zefa), 23 (Bettmann), 37 (Judy Griesedieck), 40 (Colin McPherson); ©DK Images **p. 33**; ©Getty Images **pp. 6** (Dorling Kindersley/Matthew Ward), 16 (Dorling Kindersley/Andy Crawford), 42 (William B. Plowman); ©Hungry Eye p. 24 (Greg English); ©istockphoto, background image; ©Peter Evans **p. 6**; ©Photolibrary **p. 6** (Garden Picture Library/Frederic Didillon); ©Science Photo Library **pp. 4** (Makoto Iwafuji/Eurelios), 9 (Manfred Kage), 10, 21, 46 (Eye of Science), 14 (Adrian Thomas), 15, 28 (Ian Hooton), 19 (Tek Image), 20 (Dr. Tim Evans), 29 (Patrick Dumas/Eurelios), 30 (David McCarthy), 32 (Bill Barksdale/AgstockUSA), 34 (Geoff Kidd), 35 (Mauro Fermariello), 36 (Medi-Mation), 38 (Helen McArdle), 43 (Ria Novosti), 45 (Professor Miodrag Stojkovic), 44 (Claude Nuridsany & Marie Perennou), **47** (Ron Weiss), 7. Cover image reproduced with permission of ©Corbis (Rick Miller). Background design reproduced with permission of © Istockphoto.

The publishers would like to thank Michael J. Reiss for his invaluable assistance in the preparation of this book.

Every effort has been made to contact copyright holders of any material reproduced in this book. Any omissions will be rectified in subsequent printings if notice is given to the publishers.

Disclaimer
All the Internet addresses (URLs) given in this book were valid at the time of going to press. However, due to the dynamic nature of the Internet, some addresses may have changed, or sites may have ceased to exist since publication. While the author and publishers regret any inconvenience this may cause readers, no responsibility for any such changes can be accepted by either the author or the publishers.

Contents

Always in the News....................................4

Genes and Characteristics.....................10

DNA and Proteins 16

Engineering Techniques.....................22

Genetic Technology in Medicine 26

GM Foods 32

Cloning Research 38

Genetics in the Future 44

Facts and Figures............................*48*

Find Out More *52*

Glossary..*54*

Index *56*

Some words are printed in bold, **like this**. You can find out what they mean in the glossary.

Always in the News

Genetic technology always seems to be in the news. There are stories about scientists making **clones** (exact copies) of animals, or producing mice that glow green or have an ear on their back. There are stories about plants being genetically modified (changed) to resist insect pests or produce medical drugs. There are stories of genetically modified **microbes** that can make gasoline or diesel fuel. And there are stories about scientists using human **embryos** to carry out research into new medical treatments. There have even been stories recently of scientists producing artificial life.

This baby mouse glows in the dark because of a gene added to the egg cell from which the mouse grew. The gene makes the mouse skin cells produce a glow-in-the-dark chemical.

CUTTING EDGE: GLOWING GREEN MICE

In 1999 Professor Anthony Perry and his team at the University of Hawaii used genetic engineering, in which a new gene is inserted into a plant or animal, to produce mice that glowed green. These glow-in-the-dark mice were made by inserting a jellyfish **gene** into the mice. The researchers were not trying to create novelty pets—there was a good reason for making the mice glow in the dark. Perry and his team found that if the jellyfish gene was inserted along with another, useful gene, it would act as a marker. Scientists would be able to tell easily whether the gene insertion had been successful, simply by checking whether the animal or plant glowed in the dark.

Genetic engineering is used today for many different purposes. As you will see, it is used to make medicines, increase the yield of food crops, protect plants from disease, and in many other ways.

Frankenstein science?

Sometimes the news stories suggest that genetic technology is a good thing that produces benefits for everyone. At other times, it seems that genetic technology is "Frankenstein science"—that genetic experiments are unethical (morally wrong) and the discoveries being made could be dangerous.

It is hard for most people to make their own decisions about these kinds of stories. Often, people do not know enough about genetic technology to decide whether some new development is good or bad. But it is actually very important that as many people as possible have informed opinions about these things. This is because genetic technology is an area of science in which new discoveries can have big effects on people's lives. For example, a new kind of genetically modified (GM) plant could make it possible to produce more food in places where people can hardly grow enough food to survive. But you will also need to understand the arguments against genetic modification before deciding whether the benefits of a GM plant outweigh any drawbacks.

Making good decisions

Scientists do not make decisions about genetic technology by themselves. Governments make rules about what research can be done and what products can be used. Ordinary people also have a part to play, because they are affected, too. If you understand the science, you can form clear opinions and help society make decisions that produce more benefits than drawbacks. This is why science matters.

A scientist holds two tomatoes: the one on the right has been genetically modified to grow bigger.

What is genetics?

Genetics is the science of heredity—the way the characteristics of a living thing are passed on from one generation to the next. It includes why people are different from each other, and why they differ from other animals and from plants.

If you look around your class or school, you will see that people are different from each other. For example:

- Some people are taller than others.
- Some people have blue eyes, while others have brown or gray eyes.
- Some people are good at music, while others are good at sports.

Some of this **variation** between people is caused by the genetic differences between them.

Genetics is also responsible for the differences between human beings and other **species** (kinds) of living things. It is because of genetics that dogs give birth to puppies, while acorns produce oak trees.

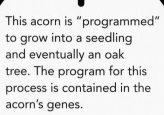

This acorn is "programmed" to grow into a seedling and eventually an oak tree. The program for this process is contained in the acorn's genes.

The beginnings of genetics

Genetics did not really exist as a science until about 100 years ago. This does not mean that no one knew anything about genetics and heredity before this time. Farmers trying to produce better **livestock** or more productive kinds of crops, or breeders trying to produce better racehorses, all knew a lot about how heredity worked in their area. Plant breeders, for example, developed new varieties of plants that produced better crops than the original varieties.

Gregor Mendel was a monk who lived in the Austrian Empire (now the Czech Republic). He carried out a large number of experiments on pea plants to try to understand how characteristics were passed from one generation to the next. His results, and the conclusions that he drew from the results, became the basis for the science of genetics.

Mendel published the results of his experiments in 1865. However, at the time, no one really understood the significance of his work. In fact, Mendel had discovered some basic rules about heredity.

CASE STUDY

Gregor Mendel (1822–1884)

Gregor Mendel (right) spent many years studying how pea plants bred together. If plants with white flowers were bred with ones with purple flowers, what was the result? What happened if the offspring (the hybrid plants) were bred again?

Mendel's experiments with pea plants revealed some important things about inheritance. He found that when simple characteristics, such as flower color, were passed on, they did not mix like paint. For example, a purple-flower plant and a white-flower plant did not produce a plant with light purple flowers. Mendel found that inheritance worked more like colored balls picked out of a bag—you got either one color of ball or another color (so, either a purple flower or a white one).

Rediscovered

In 1900 three European botanists (plant scientists) rediscovered Mendel's work. Advances in the study of heredity and in other biological sciences meant that scientists could now understand the significance of Mendel's research. This was the beginning of genetics as a modern science. For the next 50 years, geneticists were partly concerned with figuring out what the genetic material was. They knew that every animal or plant begins as a single **cell**, too small to be seen without a microscope (see box). The genetic substance had to be in this cell. But what exactly was it?

By the 1940s, scientists were fairly sure that the genetic material was a substance in the cell nucleus, which they called nucleic acid. A great research effort was made to figure out its structure. Then, in 1953, British scientist Francis Crick and U.S. scientist James Watson showed that the substance had a long, double-spiral structure (see Chapter 3). They called the genetic material **DNA** (deoxyribonucleic acid).

THE SCIENCE YOU LEARN: CELLS AND DEVELOPMENT

The human body is made up of billions and billions of tiny units known as cells. Each cell has a thin membrane around the outside. Inside the cell is a mixture of water, chemicals, and structures called **organelles**. The biggest organelle in the cell is the nucleus.

Every human, animal, and plant begins life as a single cell, called a **zygote**. This is the cell produced when genetic material from the male combines with the egg cell of the female. This zygote therefore contains material from both its mother and its father.

The zygote grows into a new animal or plant by dividing in two. Each of the two new cells has the same genetic information as the parent cell. The two cells divide to form four cells, the four cells produce eight cells, and so on. Every time the cell divides, the genetic material is copied.

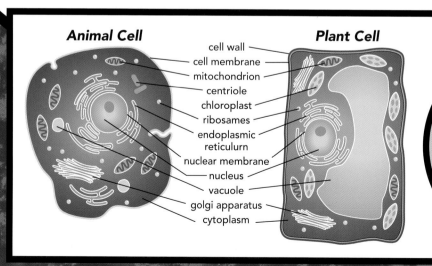

Animal Cell

Plant Cell

cell wall
cell membrane
mitochondrion
centriole
chloroplast
ribosames
endoplasmic
reticulurn
nuclear membrane
nucleus
vacuole
golgi apparatus
cytoplasm

This diagram shows the similarities and differences between an animal cell and a plant cell.

Toward genetic technology

The structure of DNA gave scientists strong clues as to how DNA reproduced itself, and how the genetic information in the DNA affected the characteristics of the person or other living thing. Over the next 20 years or so, the details of how genes normally operate were discovered.

Scientists had to develop many new techniques to unravel the secrets of genetics. From the 1970s onward, they began to realize that they could use these techniques to make changes to the genetic material, rather than just studying it. This was the beginning of genetic engineering and other kinds of genetic technology.

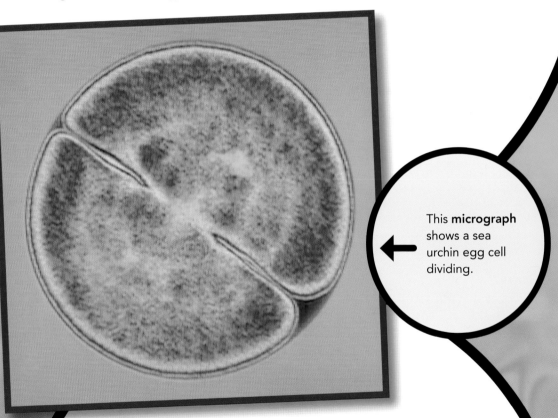

This **micrograph** shows a sea urchin egg cell dividing.

Genetic technology: For and against

In this book you will see how the genetics you learn in school has formed the basis for the many different ways that genetic technology is used today. By the end of the book you should know enough to form your own opinions. Is genetic technology "Frankenstein science," or does it benefit the world? Or is the truth more complex than that? Does genetic technology have both benefits and drawbacks that have to be balanced against each other?

Genes and Characteristics

You learned on page 8 that genetic material is called DNA and is found in the nucleus of the cell. Animals and plants get their DNA from their parents—half from the mother and half from the father. The genetic material is made up of many genes: in humans, there are about 20,000 genes. Each gene is a section of DNA with a specific job. (You will learn more about what genes do in Chapter 3.)

Two of everything

Most animals and plants reproduce sexually. This means that two sets of genes, one from the father and one from the mother, fuse to create a new animal or plant. Living things that reproduce sexually have to make special cells for reproduction, known as sex cells (gametes). In humans, female sex cells are eggs, and male sex cells are sperm. Each sex cell contains only a single set of genes, whereas all other cells in the body carry two sets of genes (see artwork opposite).

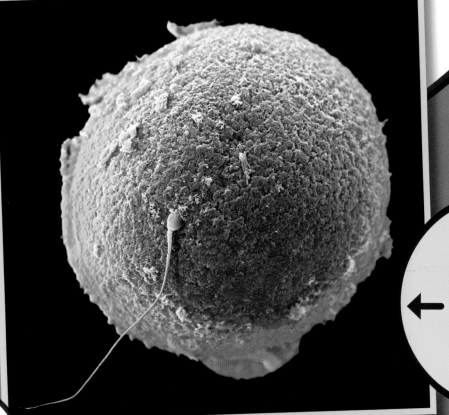

This micrograph shows a sperm cell (blue) joining with an egg cell (yellow). The head of the sperm cell contains DNA from the father.

Why are sex cells needed? It is a matter of simple math. In order to have the right amount of genetic material (two sets of genes) in each offspring cell, only one set of genes is needed from the father and only one set of genes is needed from the mother. Suppose that sex cells had two full sets of genes, like other cells. When a male and a female reproduced, each parent would pass on two sets of genes to their children. The children would then have four full sets of genes. When this second generation reproduced, each of them would pass on four sets of genes to their children—eight sets in all. And so it would go on, with the number of gene sets doubling in each generation. After 18 generations, there would be over one million copies of each gene in a cell!

Alleles

Each of a human being's 20,000 or so genes comes in two or more different "flavors," known as **alleles**. These different versions of the same gene are what make individuals different from each other. As you have just discovered, each person has two copies of each gene, one from each parent. In some cases you will get the same allele from each parent, but at other times you will have two different alleles. What happens then?

Chromosomes from both male and female sex cells are transferred so that the offspring shares chromosomes from each of its parents.

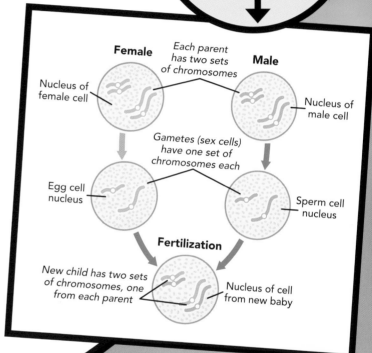

Female — Each parent has two sets of chromosomes — Male

Nucleus of female cell

Nucleus of male cell

Gametes (sex cells) have one set of chromosomes each

Egg cell nucleus

Sperm cell nucleus

Fertilization

New child has two sets of chromosomes, one from each parent

Nucleus of cell from new baby

CUTTING EDGE: EYE COLOR

Until recently, a number of different genes were thought to be associated with eye color. However, research by Dr. Rick Sturm and Dr. Tony Frukadis from the University of Queensland, Australia, found that most eye colors can be explained by changes to the DNA near a gene named OCA2 on chromosome 15. This gene produces a pigment (a colored chemical) that helps to give hair, skin, and eyes their basic color. Changes in three different places around the gene to three single bases, all located close to the OCA2 gene, can explain nearly three-quarters of the different eye colors seen in humans.

Mixing alleles

In his experiments with pea plants, Gregor Mendel found that when a purple-flower plant was crossed with a white-flower plant, all the offspring (hybrid plants) had purple flowers. Flower color in pea plants is the result of just one gene, which has a white allele and a purple allele. From Mendel's experiments, it seems that when a plant has one white-flower allele and one purple-flower allele, its flowers are purple. Only the purple-flower allele "shows" in the plant; the white-flower allele is "hidden." Mendel said that the allele (Mendel called this a "trait") for purple flowers was **dominant** over the white-flower trait, which he called **recessive**.

This characteristic of one allele being dominant over another is true for many genes. Humans, for instance, can have either free or attached earlobes (see box below).

IN YOUR HOME: GENE SURVEY

Take a look at your earlobes. How do they join to your head? Do they join right to the side of the head? Or is there a part of the lobe at the bottom that is "free"? Fixed (f) and free (F) earlobes are two alleles of the same gene, and free earlobes are dominant over fixed ones. So, people who have the alleles FF or Ff have free earlobes. Only people with two f alleles have attached earlobes.

Now take a survey of your relatives. Draw up a "family tree" showing whose earlobes are fixed and whose are free. Can you deduce anything about which alleles your relatives have?

Mendelian genetics

Now that you know about genes and alleles, you can understand what happens when a white-flower and a purple-flower pea plant breed together to produce purple-flowered plants.

- The purple-flower plant has two purple-flower alleles, which can be written "PP." A capital "P" is used for the purple-flower allele because it is dominant.

- The white-flower pea plant has two white-flower alleles, pp. These alleles are written in small letters because they are recessive.
- The offspring of the purple- and white-flowered parents get one allele from each parent, so all plants have one P allele and one p allele. The P allele is dominant, so the plants have purple flowers.

But what happens if the hybrid (Pp) plants breed together? Mendel tried this as part of his experiments, too. Over hundreds of different trials, he found that roughly three-quarters of the second-generation plants were purple-flowered and one-quarter had white flowers. Why was this? Although Mendel did not know about genes, he explained his findings in terms of "traits" for different characteristics. Today, we call his traits alleles. All the plants have the alleles Pp for flower color. If two Pp plants breed together, then four different pairings can result: PP, Pp, pP, and pp. The first three combinations all produce plants with purple flowers, since they all contain the dominant P allele. The last combination, pp, contains only the recessive p allele, resulting in plants with white flowers. So, as in Mendel's experiment, three-quarters of the plants are likely to have purple flowers, and one-quarter to have white flowers.

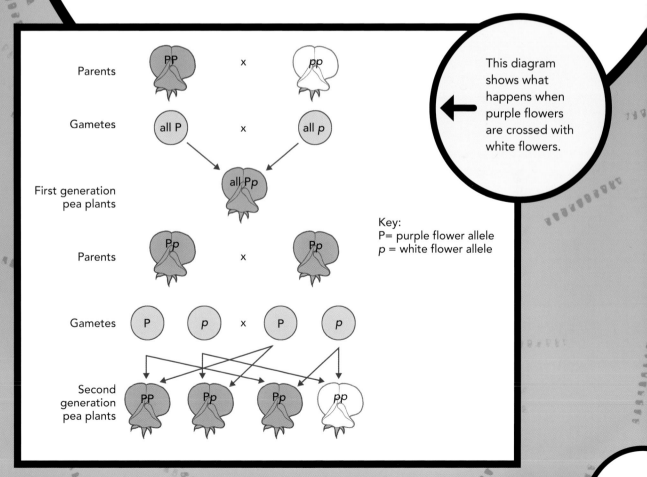

Parents   PP   x   pp

Gametes   all P   x   all p

First generation pea plants   all Pp

Parents   Pp   x   Pp

Gametes   P   p   x   P   p

Second generation pea plants   PP   Pp   Pp   pp

This diagram shows what happens when purple flowers are crossed with white flowers.

Key:
P = purple flower allele
p = white flower allele

13

Not always like Mendel

When Mendel did his genetic studies, he was lucky in his choice of plant and in the characteristics he chose to study. Each characteristic he studied (for instance, dwarf or normal-sized plants, wrinkled or smooth peas, white or purple flowers) was controlled by a single gene with two different alleles.

However, in many cases, figuring out the connections between genes and characteristics is more complex. There can be many reasons for this.

Sometimes a gene may have more than two alleles. For instance, human ABO blood types are controlled by a gene that has three different alleles. Four different blood types result from combinations of these three alleles—types A, B, AB, and O. Understanding these blood types has made it possible for doctors to match people with compatible types of blood, so that someone who has lost a lot of blood can be given a blood transfusion (blood from another person).

In other cases, one gene is not dominant over another. If two alleles both "show" to the same extent, this is called incomplete dominance. For example, snapdragons have a flower-color gene that has two alleles, one for red flowers and one for white flowers. If a plant has one red-flower gene and one white-flower gene, both genes are expressed, and the plant has pink flowers.

In the snapdragon case, the red-color gene codes for producing red pigment, and the white-flower gene produces no pigment. In a third kind of dominance known as co-dominance, different genes code for different proteins, which are produced independently. The A and B allele blood types in the ABO blood system are co-dominant, because each codes for a different protein on the surface of blood cells. (The O allele does not code for any cell surface protein.)

Snapdragon flowers can be red, white, or pink, depending on the combination of alleles a particular plant contains.

Most often, the main reason why it is difficult to trace the relationship between genes and characteristics is because a characteristic is not due to a single gene. Most human characteristics, such as hair or skin color, height, and body shape, result from the combined effects of a number of genes. In such cases, it is hard to make connections between individual genes and the effects they produce.

THE SCIENCE YOU LEARN: PUNNETT SQUARES

Punnett squares are a kind of diagram in which it is easy to write down all the possible combinations of alleles a person can inherit from his or her parents. One parent's alleles are written along the top edge of the square, while the other parent's alleles are written down the left-hand side. The different possible combinations are written in the square itself. So, a Punnett square for the combination of Mendel's purple-flowered peas with the white-flowered peas would look like this:

	Female alleles	
	P	P
Male alleles p	pP	pP
p	pP	pP

For a second-generation cross between the first-generation plants, the Punnett square would be as follows:

	Female alleles	
	P	p
Male alleles P	PP	Pp
p	pP	pp

The color of our eyes, hair, and skin is the result of the combined effect of a number of genes.

DNA and Proteins

By the 1940s, scientists had a broad, general idea of how genetic information was passed on from generation to generation. But how exactly did the whole process work? The breakthrough that led to the understanding of what genes are and how they work came when the structure of DNA was discovered in 1953.

The structure of DNA

By the early 1950s, scientists were sure that the genetic material was the nucleic acid in the nucleus of cells. Groups of researchers around the world raced to discover the exact structure of this material. Two groups of scientists, in London and Cambridge, England, eventually cracked the problem in 1953. The London scientists, Maurice Wilkins and Rosalind Franklin, made crystals of the material and took special X-ray photographs of them. Two scientists in Cambridge, James Watson and Francis Crick, figured out the final structure using the crystal photographs and other evidence.

In the scientific paper describing the structure of the genetic material, Watson and Crick called it DNA (deoxyribonucleic acid). They showed that each **molecule** of DNA consists of two enormously long chains of **atoms**, joined together in many places along their length. The double chain is twisted in a helix, or spiral.

Each of the chains of DNA is made up of millions of repeating sub-units, known as **nucleotide bases** (often just called bases). The entire molecule is made from just four different bases, known as A, T, G, and C (adenine, thymine, guanine, and cytosine).

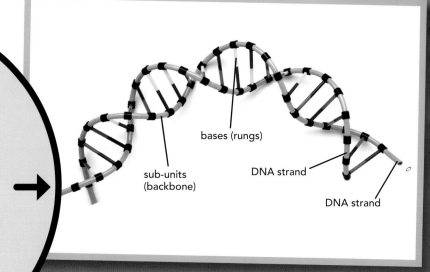

This is a simple model of the structure of DNA. It is made up of two DNA strands, joined to form a twisted "ladder." Each strand has a "backbone" of repeating sub-units: these are the sides of the ladder. Sticking out from the backbone of each strand are the different bases. These join the two strands together, forming the "rungs" of the ladder.

bases (rungs)

sub-units (backbone)

DNA strand

DNA strand

The genetic code

How can DNA, which is made from just four different sub-units, contain a code for making proteins, which have 20 different sub-units? Researchers found that every three bases along the length of a gene codes for one amino acid. Each triplet of bases is known as a **codon**. So, for example, the sequence GCC is the codon for the amino acid alanine, while CGU is the codon for arginine.

There are 64 different ways to write down a three-letter sequence of four different bases, but only 20 amino acids to code for. So, some amino acids therefore can be coded by several different codons. There is also a codon (AUG) meaning "start," and several for "stop."

Evidence for the genetic code came from an experiment carried out by Francis Crick and co-workers in 1961. They caused a series of mutations (changes) to DNA from a **virus**. Each mutation involved deleting one base from the DNA. When they made one deletion, the virus produced a protein that was damaged and did not work. When they made two deletions, the protein was still damaged. However, when they made three deletions, the virus once again produced a working protein. The illustration below shows how this worked.

These diagrams use words to illustrate how Crick's experiment showed that the genetic code worked in units of three. →

A Before there are any mutations, the gene produces a correct protein. The sequence of bases "makes sense."

THE/BIG/BEE/ATE/ONE/COW

B The mutation inserts an extra base at the start of the sequence. Now when it is divided into threes, the sequence does not make sense.

HTH/EBI/GBE/EAT/EON/ECO/W

C A second base is inserted at the start of the sequence. The sequence still does not make any sense.

GHT/HEB/IGB/EEA/TEO/NEC/OW

D A third base is now inserted. The first triplet does not make sense, but the rest of the sequence can be read again.

CGH/THE/BIG/BEE/ATE/ONE/COW

From DNA to proteins

Once scientists had figured out the structure of DNA, they began to unravel the mystery of how it worked. Over the next 10 or 12 years, researchers showed that DNA was a template or recipe book of instructions for making proteins.

Proteins are the most important substances in the day-to-day running of cells. Protein molecules are long chains made of sub-units called amino acids. While DNA has just four different sub-units, there are 20 different amino acids. The order of amino acids in the protein chain has to be correct for each protein to work properly. This is why protein recipes are stored in the DNA.

Researchers found that proteins were made outside the nucleus, but DNA never leaves the nucleus. The instructions for making a particular protein are instead copied to another kind of nucleic acid, called messenger RNA. The messenger RNA travels out of the nucleus to organelles called **ribosomes**, where the instructions coded along its length are turned into a protein.

This diagram shows how instructions are passed from the nucleus to create a new protein.

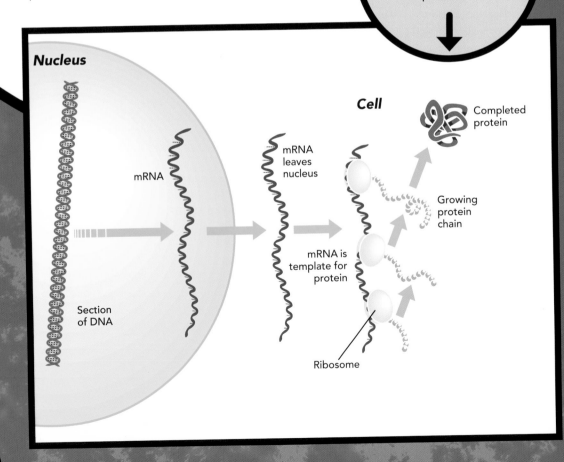

Nucleus

mRNA

Section of DNA

mRNA leaves nucleus

mRNA is template for protein

Cell

Completed protein

Growing protein chain

Ribosome

A geneticist examines the genetic sequence in a strand of DNA.

SCIENCE IN YOUR TOWN: DNA FINGERPRINTING

Most sections of DNA are the same for every human being, but there are some sections that are different. These differences make every person a unique individual. In DNA "fingerprinting," many short pieces of DNA from regions that vary between individuals are compared. If two samples have the same pattern of DNA pieces, they are almost definitely from the same person. The chance of two people having the same DNA "fingerprint," unless they are identical twins, is extremely low.

DNA fingerprinting is a very important tool in helping police catch criminals. Tiny amounts of blood, saliva, or skin from a crime scene can be tested for its DNA fingerprint. The results can be compared with those the police already hold in a huge DNA database containing the unique DNA record of millions of individuals. The police can also test the DNA of anybody suspected of the crime. If two DNA fingerprints match, this is good evidence that the person was at the crime scene.

The meaning of a gene

Once the mechanism for making proteins from DNA had been unraveled, scientists could see the relationship between genes and DNA. Each DNA molecule has many genes along its length. The genes are sections of DNA that code for a particular protein, or sometimes a group of proteins.

A protein (or group of proteins) coded for by a particular gene has a specific job in the body. If the protein has an obvious effect—for example, producing a pigment that affects the color of a flower—then the gene can be related to a particular characteristic of the living thing. Most proteins have much less clear effects, however, so it is hard to relate particular characteristics to their underlying genes.

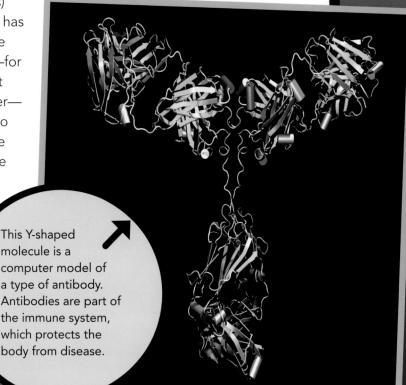

This Y-shaped molecule is a computer model of a type of antibody. Antibodies are part of the immune system, which protects the body from disease.

 THE SCIENCE YOU LEARN: PROTEINS

Proteins have a variety of jobs in the body. There are two basic kinds: fibrous and globular.

- Fibrous proteins have a repetitive pattern of amino acid sub-units, and, as their name suggests, they usually form fibers. This kind of protein plays an important part in the structure of the body. Collagen, which is found in bones, and keratin, which is part of nails and hair, are both fibrous proteins.
- Globular proteins do not have repeating patterns of amino acids. In globular proteins, the protein chain folds itself up into a particular shape. Most globular proteins are **enzymes**. Enzymes are very important to the body. They are the substances that make life processes happen, second by second. Enzymes control the thousands of chemical reactions that happen in the body. Each enzyme speeds up a particular chemical reaction thousands of times.

Sickle cell disease

Sometimes researchers have been able to link a particular gene to a protein and to an effect on the body. Often, this has been possible because of a genetic disorder that affects just one gene in the body. An example is sickle cell disease. This is a disorder in which some of the red blood cells change shape and become like sickles (crescent shaped). The sickle cells tend to stick to each other, which can block up blood vessels.

People with sickle cell disease are most often found in parts of Africa, the Middle East, and southern Asia, where the disease **malaria** is common. This is because there is a link between sickle cell disease and malaria. People who are **heterozygous** (have only one allele) for the disease show almost no symptoms (signs of the disease), but are protected against malaria. Those who are **homozygous** (have two sickle alleles) for the disease suffer from chest pain, difficulty in breathing, and damage to organs such as the spleen and the lungs.

Genetic studies have helped doctors understand sickle cell disease. In 1949 the U.S. scientist Linus Pauling showed that sickle cell disease is the result of an abnormality in the protein **hemoglobin**. This is caused by an error in the genes that code for hemoglobin. There are large amounts of hemoglobin in all red blood cells. It is essential for carrying oxygen from the lungs to the parts of the body where it is needed.

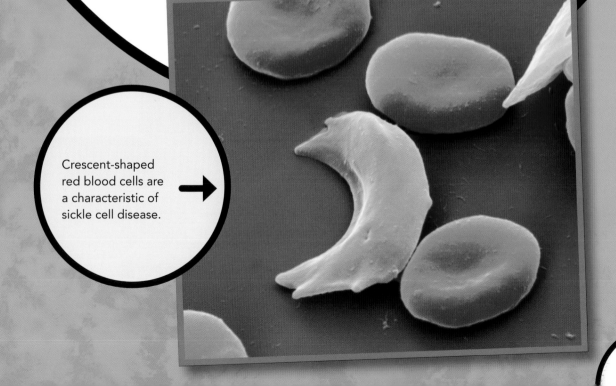

Crescent-shaped red blood cells are a characteristic of sickle cell disease.

Engineering Techniques

In their early investigations, scientists had to develop a whole range of techniques for dealing with DNA and other kinds of genetic material. Many of these methods involved using enzymes.

DNA polymerases

One of the first enzymes that researchers managed to isolate from cells was called DNA polymerase. This is an important enzyme for copying DNA.

Before a cell can divide, it has to copy its entire DNA, so that both cells have a full set of genetic material. The way this happens naturally is:

- In one area, the two strands that make up the DNA chain untwist and "unzip" from each other.
- On each of the two single DNA strands, DNA polymerase goes into action and makes a new DNA strand that is complementary to the original strand. This means that each nucleotide base in the new strand is a match for the original strand and can connect to it.
- The result is two molecules of DNA instead of one.

In the laboratory, scientists use DNA polymerase to copy strands of DNA that they want to study. They use gentle heat to unzip the two DNA strands, then add DNA polymerase and supplies of nucleotide bases to the mixture. Another useful enzyme is DNA ligase. This enzyme is used in the cell to repair damaged DNA. It can join together two pieces of double-stranded DNA that have been cut through.

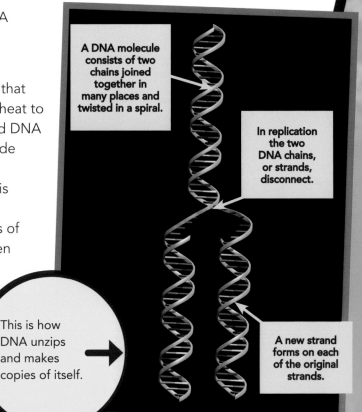

A DNA molecule consists of two chains joined together in many places and twisted in a spiral.

In replication the two DNA chains, or strands, disconnect.

This is how DNA unzips and makes copies of itself.

A new strand forms on each of the original strands.

DNA sequences

In the 1970s, British scientist Frederick Sanger (below) developed a method for **sequencing** a piece of DNA (discovering the exact order of bases along its length). In 1977 he sequenced the DNA of a virus that had over 5,000 bases. Later, he found the sequence of another virus with nearly 50,000 bases. He won a Nobel Prize for this work in 1980.

In the 1990s, Sanger's technique was made much faster by the use of automated machines and computers. A worldwide research effort began, called the Human Genome Project, that aimed to sequence the human **genome** (all the human genetic material). In total, the human genome contains over three billion bases. These are stored in human cells, along with 23 pairs of chromosomes. Each chromosome is a single, huge DNA molecule that is millions of bases long.

Despite the enormous size of the task, the Human Genome Project produced the first complete map of human DNA in 2003. Its next aim is to identify every single gene in the human genome. (Experts estimate there are about 22,000 to 23,000 genes.)

PCR

In 1983 the U.S. scientist Kary Mullis (see box below) invented a technique using DNA polymerase that made it possible to make millions of copies of a DNA strand. The process is called the polymerase chain reaction, or PCR. PCR can be used to "amplify" a tiny amount of DNA as much as 100 billion times.

PCR has been extremely useful in DNA research. It is also useful in other ways. PCR can be used to detect pathogens (**bacteria** or other microbes that cause disease)—for example, in samples of food after an outbreak of food poisoning. By using PCR in DNA fingerprinting, it becomes possible to get DNA fingerprints from the tiniest sample simply by "amplifying" it first.

CASE STUDY

Kary Mullis

One spring night in 1983, Kary Mullis (right) was driving along the highway in northern California. He was thinking about his work, which involved copying small pieces of DNA. As he drove, he thought of a way to make lots of copies of a piece of DNA, starting from just one DNA molecule. Mullis's idea later became PCR (the polymerase chain reaction).

The essence of Mullis's idea was to use repeated changes of temperature to set up a cycle of events. DNA was split into single chains, then copied, then split into single chains again, and so on. At first there was a problem that slowed everything down. Every time the mixture was heated to make the DNA chains split apart, the DNA polymerase that did the copying became damaged. New DNA polymerase had to be added each time. Then, Mullis discovered a type of DNA polymerase that could survive being heated and cooled again. This meant that PCR could be done automatically, by a machine.

Selective scissors

Another set of enzymes that have proved extremely useful for genetic research are called restriction enzymes. Restriction enzymes were originally found in bacteria. Bacteria use restriction enzymes to defend themselves against viruses, by cutting up viral DNA. Restriction enzymes act like scissors, cutting through both strands of a DNA molecule and breaking it up into smaller pieces. However, restriction enzymes do not cut the DNA in a random place. Each restriction enzyme cuts the DNA only where there is a particular sequence of bases in the chain. Because they cut DNA in specific places, restriction enzymes produce fragments of the same size if they are used to cut up identical strands of DNA. This fact is the basis for DNA fingerprinting.

Not all restriction enzymes make a clean cut straight across the two DNA chains. In some cases, the enzyme cuts one strand in one place and the other strand in a different place. The result is two sections of DNA with "sticky ends"—short sections of single-stranded DNA at the ends. These sticky ends can be very useful, because a piece of DNA with sticky ends will join up with another DNA fragment with similar sticky ends. The two DNA pieces can then be joined permanently using the enzyme DNA ligase (see the diagram).

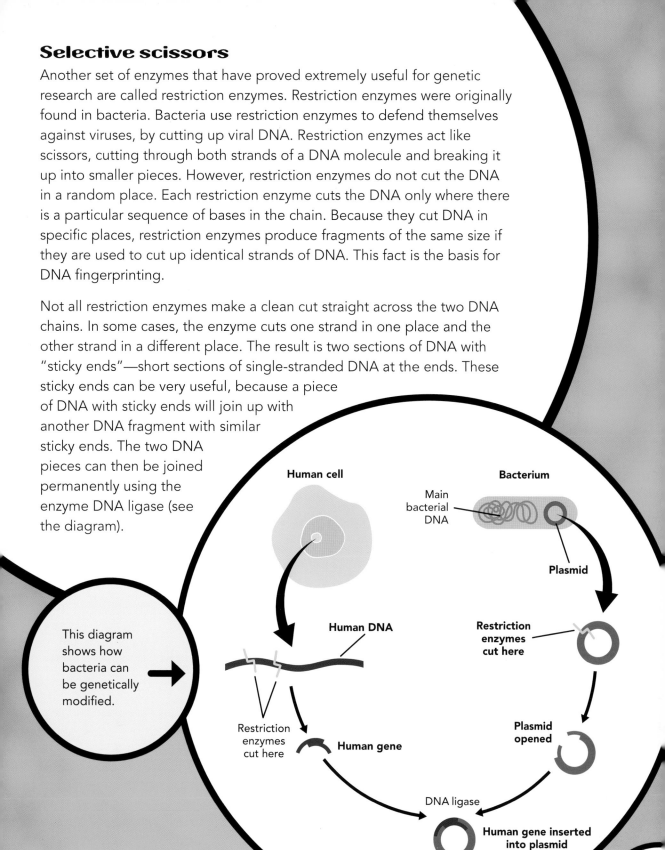

This diagram shows how bacteria can be genetically modified.

Human cell

Bacterium

Main bacterial DNA

Plasmid

Human DNA

Restriction enzymes cut here

Restriction enzymes cut here

Human gene

Plasmid opened

DNA ligase

Human gene inserted into plasmid

Genetic Technology in Medicine

One of the first ways that genetic technology was used was to produce medicines. To do this, researchers genetically modified bacteria with human genes.

Modifying bacteria

How did scientists manage to put human genes into bacteria? The main tools in the process are restriction enzymes, DNA ligase, and circular pieces of DNA from bacteria known as **plasmids**. Plasmids are short, circular pieces of DNA produced by bacteria that are separate from the bacteria's main genetic material. They are genetic add-ons that can help the bacteria survive in certain conditions. Scientists have found that bacteria will take up plasmids from their surroundings if conditions are right. Plasmids are therefore a good way of introducing genetic material into bacteria.

The first step in the process is to separate plasmids from bacteria and cut them open. This is done using a restriction enzyme that produces sticky ends (see page 25). Next, the section of human DNA containing the desired gene has to be cut using the same restriction enzyme, so that it has the same sticky ends as the plasmid. If the opened plasmids and the sections of human DNA are now mixed, the sticky ends of both pieces of DNA will stick together, creating a plasmid with a human gene added into it. The joins are made permanent using DNA ligase. If the engineered plasmids are now mixed with a colony of bacteria, some of the bacteria will take up the plasmids. The bacteria that do this will start producing the protein that is coded for by the human DNA in the plasmid.

Medicines from bacteria

Some medicines are produced by genetically modified bacteria. Colonies of the bacteria are fermented to make them grow and multiply. This means giving the bacteria food, keeping them warm, and allowing them to grow in the absence of air. Under these conditions, the genetically modified microbes produce vaccines or other medicines. For example, two substances that are important in blood clotting are made in this way. These are used to treat people with hemophilia, an illness in which the blood will not clot.

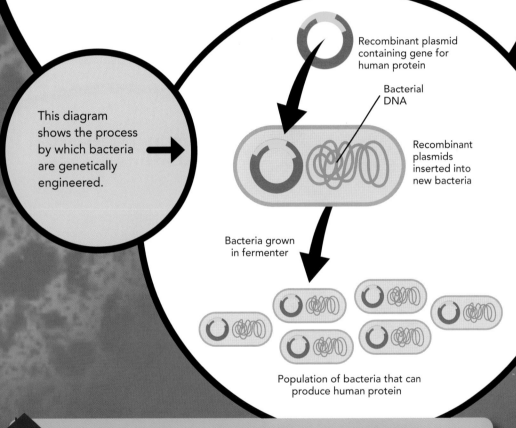

This diagram shows the process by which bacteria are genetically engineered.

Recombinant plasmid containing gene for human protein

Bacterial DNA

Recombinant plasmids inserted into new bacteria

Bacteria grown in fermenter

Population of bacteria that can produce human protein

IN YOUR HOME: FERMENTING MICROBES

People have been fermenting microbes, such as **yeast**, for different purposes for thousands of years. You probably have some **fermentation** products in your kitchen right now.

- Special bacteria are used for yogurt-making. They produce acid when they are fermented, which gives the yogurt a sharp taste.
- Soy sauce is made by fermenting soybeans with a mixture of different microbes.
- Beer and wine are also fermented products.

Human insulin

The first bacteria to be genetically modified for medical use contained a gene to make human insulin. Insulin is a substance that is essential for maintaining the level of sugar (glucose) in the blood. People with the illness diabetes cannot make enough insulin, so they cannot control their blood sugar levels. If untreated, diabetes can have severe effects on the body, including sight loss, kidney disease, and, eventually, death.

Until human insulin was made using GM bacteria, patients with diabetes were treated with insulin obtained from pigs or cattle. For most people this worked fine, but some patients had a bad allergic reaction to pig or bovine insulin. Today, human insulin is made from GM bacteria in fermenters. Large amounts of insulin can be made cheaply, and human insulin works better than pig or bovine insulin.

Spotting genetic disorders

Another way that genetic technology has been used in medicine is to test for genetic disorders. A genetic disorder is an illness caused by a small mistake in a person's genetic makeup. Sickle cell disease (see page 21) is a genetic disorder.

There are about 10,000 known genetic disorders in humans. Most of them are very rare, but a few are relatively common. Cystic fibrosis, for example, is a disease that causes symptoms such as repeated chest infections and poor growth. About 30,000 Americans have cystic fibrosis. Other genetic disorders include Down's syndrome and muscular dystrophy.

All insulin used by diabetics today comes from genetically engineered bacteria.

In many cases, scientists have discovered through research which gene is affected in a genetic disorder. This makes it possible to test young babies, or even embryos still in the mother's womb, to find out if they have a particular genetic disorder.

Modern tests for a genetic disorder are usually done using a DNA array. This is a small, plastic square that has hundreds, or even thousands, of different pieces of DNA attached to it. Each DNA piece is a probe that can detect a specific genetic disorder. By using DNA arrays, doctors can test for hundreds or even thousands of genetic disorders at once.

A researcher examines a DNA microarray. The glass slide contains hundreds of different samples of DNA, each one "tagged" with a dye that fluoresces (shines) in ultraviolet light.

Knock-out mice

One of the main ways that scientists have been able to learn about genetic disorders has been to study them in other animals, especially mice. Knock-out mice have been very important for showing which genes cause a particular disease. These are mice in which a specific gene has been disabled, or knocked out. Knocking out the activity of a gene provides valuable clues about what that gene normally does. For example, the p53 knock-out mouse is missing gene p53. Mice missing this gene are far more likely than normal mice to develop some kinds of cancer. This information has led to the discovery that the p53 gene produces a protein that protects against some forms of cancer.

Humans share many genes with mice. So, studying the characteristics of knock-out mice gives researchers information that can be used to better understand similar problems in humans. Studying knock-out mice has shown how genes can affect cancer, obesity, heart disease, diabetes, aging, and many kinds of illness.

CASE STUDY

Gene therapy

Spotting a genetic disorder at an early stage can help doctors to treat the disorder. For example, cystic fibrosis, one of the most common genetic disorders, cannot be cured at present, but doctors can give patients **antibiotics** and other medicines to help treat the symptoms. Perhaps the only way to truly cure a genetic disorder would be to replace the damaged gene in the patient's DNA with an undamaged copy. This is known as **gene therapy**.

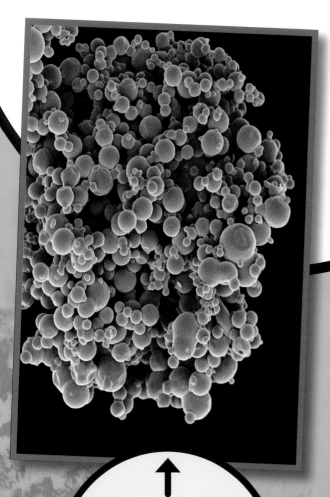

Liposomes are artificially made capsules that have an outer membrane similar to a cell membrane. Researchers are experimenting with using liposomes as carriers instead of using modified viruses.

CUTTING EDGE: GENE THERAPY IN HUMANS

Gene therapy has been successfully used in animals, but it has had very mixed success in humans. In 1999 Jesse Gelsinger, a teenager with a genetic disease of the liver, died as a result of an attempt to cure his illness by gene therapy. After his death, the rules about using gene therapy in humans became tighter.

Since 2000, several children with a genetic condition that leaves them with no defenses against disease have been treated successfully by gene therapy. However, there is a risk that these children might develop leukemia (a kind of cancer of the blood). More recently, gene therapy has been used to treat a sight problem affecting the retina (the light-sensitive layer on the inside of the eye). However, it is too early to tell if this use of gene therapy has been successful.

Experiments in animals have shown that gene therapy is possible. To do this, a normal copy of the damaged gene is inserted into the DNA of a virus. The virus is then injected into the animal, either in the blood or at a target site. Viruses have built-in mechanisms for getting into cells and mixing their own DNA with that of the cell. The virus therefore acts as a "carrier" to transport the new gene into the nucleus of the animal's cells.

In one experiment in St. Louis, Missouri, newborn dogs and mice that had the genetic disease hemophilia were given a gene that produces a protein that is necessary for blood clotting. The gene cured the animals of hemophilia, and they were still healthy over one year after the treatment.

Gene therapy has been tested in humans, but there are still many problems with the technique. One worry that researchers have is that the viruses used to insert the new gene into the body could cause illness, in particular cancer. Until gene therapy has been shown to be safe and effective, it will not be widely used.

CUTTING EDGE: MAKING SPIDER SILK

Genetic engineering is not only used to make medicines. One very promising material that is already being made using GM tobacco plants and **transgenic** goats is spider silk. Spider silk is stronger than steel, but also elastic and very light. It could be used for many different applications, ranging from super-thin bulletproof vests to fishing lines. It could also be used inside the body, to replace damaged **tendons** or **ligaments**.

Transgenic animals are animals of one species that have been genetically modified with genes from another species. These transgenic goats have been modified to produce spider silk in their milk.

GM Foods

The area of genetic technology where there are perhaps the most heated debates about the benefits and risks of genetic technology is genetically modified (GM) plants. GM food crops were first grown on a large scale in 1996. Today, farmers in 16 countries, notably the United States, China, Canada, and Australia, grow genetically modified food plants. The main GM plants are soybeans, corn (maize), rapeseed, and cotton. In the European Union and Japan, GM crops are not grown commercially. However, many prepared foods sold in these countries contain low levels of GM soybeans and corn.

GM soybeans are sprayed with herbicide (weed-killer). "Roundup-ready" soy is modified to be very resistant to herbicide. Large doses can be used to kill off all weeds around the soy plants.

IN YOUR HOME: SPOTTING GM FOODS

In some countries, there are laws saying that GM foods should be labeled. However, manufacturers do not always include a GM label, and in some countries there is no need to label GM food. Soybeans and corn are the two most common GM crops. Any ingredients containing corn or soy could include at least some GM crops, although there is no way of being certain. Possible GM ingredients include:

- cornstarch, corn oil, modified starch, starch, corn syrup, dextrose
- soybeans, soy flour, soybean oil, soy sauce, soy lecithin (E322), tofu, hydrolyzed vegetable protein
- tomato paste, tomato puree, tomato sauce
- rapeseed or canola oil
- any products that say "vegetable oil," as this is probably rapeseed, canola, corn, or soybean oil
- the artificial sweetener aspartame.

Selective breeding and cloning

People have been changing plants to make them more useful to humans for thousands of years. Modern wheat, for example, is very different from the earliest wild wheat, which people in the Middle East first gathered about 10,000 to 15,000 years ago. Modern wheat has been improved by thousands of years of selective breeding. This is when farmers choose the plants that produce the most food, or are the best tasting, and grow next year's crop from these plants. Over many years, selective breeding can produce big changes.

Another way that people select the most useful plants is by cloning them (producing many plants that are genetically identical). It is possible to make clones of many plants by taking cuttings. This involves cutting off small pieces of a plant (usually new shoots) and planting them. With care, the cuttings will grow into new plants that are identical.

This photo shows different kinds of wheat. The types of wheat with long bristles are older emmer wheat. The wheat without bristles is modern triticum wheat.

Doing it more quickly

Selective breeding and cloning from cuttings are actually kinds of genetic technology. However, it can take many years to produce plants with the preferred characteristics using selective breeding. Using genetic modification, however, plants can be changed more quickly, and it is possible to add very specific new characteristics from other living things.

What are the main GM crops?

All the GM crops currently grown on a large scale are of two broad types. In one type, genes have been added that make the plant resistant to high doses of herbicide (weedkiller). Such high doses would kill any other type of plant. These kinds of crops only have to be sprayed with weedkiller once during their growth. This means that, overall, less herbicide is used, because normal crops have to be sprayed several times during their growth. Soybeans and corn are the most common crops to be modified in this way.

The second kind of widely grown GM crop is one that has been modified to produce a toxin (poison) that kills insect pests but does not affect humans or other animals. This treatment is very effective at getting rid of insects without using pesticides (insect-killing chemicals). Cotton and maize are the crops most commonly modified in this way.

This crown gall tumor was caused by *A. tumefaciens*.

How are they made?

GM crops are made in two main ways. The first of these is to use a bacterium called *Agrobacterium tumefaciens*. This bacterium causes a problem called crown gall disease in plants. It does this by infecting plants and inserting its genes into plant cells.

To use *A. tumefaciens* for genetic modification, the bacterium's own genes are first completely removed. They are then replaced with the genes that the researchers want to insert into the plant. The bacteria are then used to infect the target plants. One example of a successful GM crop is Bt cotton or Bt corn. The plants are modified with a gene called Bt. This produces a protein that is highly poisonous to insect pests. Insects that try to eat Bt plants are poisoned.

Another way of inserting genes into plants is to use a "gene gun" (left). This piece of apparatus fires DNA-coated genes directly into plant cells (in the petri dish). Some of the cells are destroyed, but others take up the genes and can be grown into new plants.

Discovering A. tumefaciens

In the 1970s, Belgian scientists Marc Van Montagu and Jeff Schell were investigating the interactions between plants and bacteria in the soil. They discovered that the bacterium A. tumefaciens could transfer genes into plant cells. Montagu and Schell developed ways to alter the bacterium so it could be used to insert other genes into plants. This enabled them to make the first GM plant—a tobacco plant that was resistant to certain pests.

Other GM crops

Plants have been genetically modified in many different ways. Some have been modified to be more resistant to decay or to grow well in a hot climate or in poor soil. Other plants have been genetically modified to be more nutritious. Golden rice is a GM crop that is rich in vitamin A. Normal rice does not contain much vitamin A, and in poor countries where rice is the **staple food**, children often suffer from vitamin A deficiency. This can cause blindness and other kinds of illness. The aim of golden rice was to make a basic food that could prevent vitamin A deficiency.

The first version of golden rice did not contain enough extra vitamin A to make it a useful alternative to normal rice. A newer version, much richer in vitamin A, has now been developed, and this is currently being scientifically tested. However, the value of golden rice is widely debated. Critics say that poor people who need the rice are deficient in a range of other vitamins and nutrients, too. They argue that the money spent developing a high-tech solution like golden rice would be better used to give these people a more balanced diet.

Vaccine code: RD300 35 40 00X

This is a computer-generated image of plants that have been genetically engineered to produce vaccines against human diseases. The method is a cheap way of producing large quantities of vaccine.

GM animals

Most GM animals are very expensive to produce and breed, so killing them for food is not economical. One kind of GM animal that is being developed for food is salmon. The salmon are modified with genes that make them grow bigger and more quickly. Scientists first produced GM salmon in the late 1990s, but the salmon have not yet passed all safety tests to allow them to be sold as food. One of the main worries is what would happen to wild salmon if GM salmon escaped. Some laboratory tests suggest that GM fish would out-compete the wild ones, but laboratory studies are not always an accurate guide to what happens in the wild.

GM salmon are capable of growing up to 30 times faster than natural salmon. ➡

Opposition to GM foods

GM crops are grown widely in North and South America and in some Asian countries. However, many people, especially in Europe, are opposed to growing any GM foods. One of the main arguments against using them is that it is almost impossible to stop the seeds of GM crops from spreading into the wild. There is concern that they could "contaminate" the plants grown on other farms or become "super weeds" that cannot be killed using herbicides.

IN YOUR HOME: GM ENZYMES

Enzymes are used in many different ways in food production. For example, when manufacturers make fruit juice, they add an enzyme to the fruit that breaks down plant **cell walls**. This makes the fruit release more juice without ruining its color. Another common enzyme turns milk into a lumpy solid (curds) and a thin liquid (whey). This is one of the first stages of cheese-making. Since 1990 many enzymes used in food production have been genetically engineered.

Cloning Research

What is the connection between a lump of baker's yeast, a swarm of summer aphids, and identical twins? The answer is that they are all clones. Clones are living things that are genetically identical to each other.

- Identical twins are clones because they both come from the same zygote (**fertilized** egg cell).
- Baker's yeast is a kind of tiny microbe. The yeast microbes reproduce by cell division (splitting in half), so all the yeast cells are genetically the same.
- In summer, female aphids lay thousands of eggs that have not been fertilized by a male. These eggs all hatch out as females. They are clones of the mother aphid because only her genes have gone into the eggs.

All the clones described above are found in nature. However, people have also found ways to make clones artificially. Scientists have been able to clone plants for many years, using techniques such as cuttings (see page 33) and **grafting** (see box to right). More recently, scientists have found ways to clone animals.

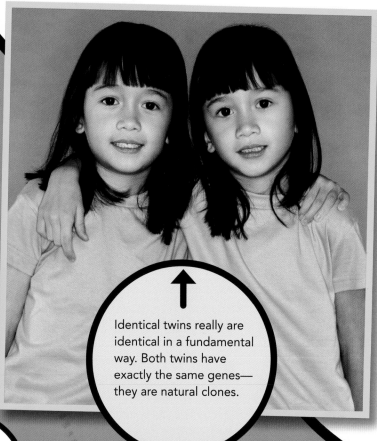

Identical twins really are identical in a fundamental way. Both twins have exactly the same genes—they are natural clones.

The first artificial clones

The earliest animals to be cloned were tadpoles. In 1952 U.S. scientists Robert Briggs and Thomas J. King successfully cloned 27 of them. They did this by taking the nuclei from the cells of a tadpole embryo (a tadpole in the very early stages of development). They then took the nuclei out of some tadpole eggs and replaced them with the nuclei from the embryo. This technique is known as **nuclear transfer**.

Many kinds of fruit are clones. For example, all the apples of a particular variety, such as Cox's or Royal Gala, are identical clones. Each apple tree of that variety is grown by grafting a small piece of one tree onto the rootstock (the roots and a small part of the trunk) of another tree. (See the diagram to the right.) Sometimes pieces of several different apple trees are grafted onto one stock, to make a tree that produces several different kinds of apple. Apple trees are grown this way to ensure that each tree produces good quality apples. If apple trees are grown from seed, more often than not they produce wild crab-apples, which are small and very sour.

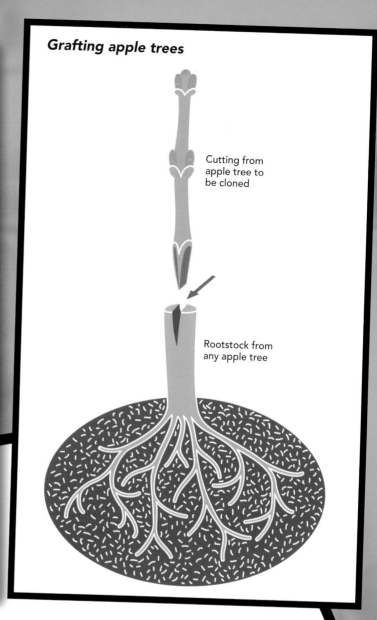

Grafting apple trees

Cutting from apple tree to be cloned

Rootstock from any apple tree

Cloning mammals

Nuclear transfer worked well for frogs and similar animals, but researchers found that it did not work for mammals. However, in the 1980s, a Danish scientist, Steen Willadsen, found a different way to clone farm animals. He cloned a cow by taking a young calf embryo and dividing it up into individual cells. He then placed each of these cells in the womb of a female cow and allowed them to develop.

After this success, Willadsen tried cloning sheep by nuclear transfer. As Briggs and King did with frogs, he took the nuclei from sheep embryo cells and put them into eggs with their nuclei removed. In 1984, after several failures, he managed to clone a sheep from embryo cells.

Dolly the sheep

In 1996 a group of researchers in Edinburgh, Scotland, led by Ian Wilmut and Keith Campbell, produced the world's first mammal cloned from adult cells. The cloned animal was a sheep named Dolly. Wilmut and Campbell used a nucleus from one adult sheep, which they put into an egg from a second sheep. The egg was then put into the womb of a third sheep to develop. Dolly lived to the age of six and produced six lambs. These were not clones, but rather ordinary lambs produced by normal reproduction.

After the success of Dolly, many other researchers cloned animals using the same technique. Many cows, goats, sheep, and pigs have been cloned. The first cat was cloned in 2001, and the first dog in 2005. Then, in 2007, scientists in Oregon produced 20 cloned monkey embryos. Monkeys are very similar genetically to humans, so this was a big breakthrough.

The successful cloning of Dolly the sheep caused a sensation in 1996. She is seen here with one of her creators, the embryologist Professor Ian Wilmut.

Cloning Dolly

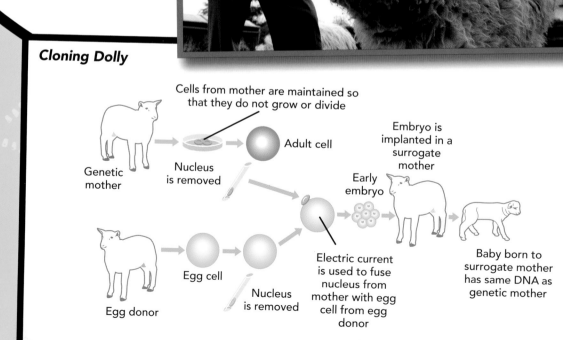

Cells from mother are maintained so that they do not grow or divide

Genetic mother

Nucleus is removed

Adult cell

Egg donor

Egg cell

Nucleus is removed

Electric current is used to fuse nucleus from mother with egg cell from egg donor

Early embryo

Embryo is implanted in a surrogate mother

Baby born to surrogate mother has same DNA as genetic mother

Why make clones?

Why do scientists want to make artificial clones? There are two main reasons.

1. To produce more individuals of an exceptional quality

Fruit farmers graft well-known varieties of apples onto trees because they want to grow apples that are good to eat. Scientists became interested in cloning in order to create transgenic animals that could produce drugs or other substances in their milk (see box below). Conservationists would perhaps be interested in cloning animals that are endangered (see page 43).

2. Stem cell research

Some people argue that the most important use for cloning would be to create special kinds of cells known as embryonic **stem cells**. These stem cells are only found in very young embryos, when the offspring is no more than a ball of cells. They have an amazing ability to turn into any kind of body **tissue**. It is thought that stem cells could be used to repair damaged organs, re-grow broken nerves, or possibly grow new organs for transplantation into a sick patient.

CASE STUDY

Mira and her sisters

In 1999 three goats named Mira were born on a farm in Massachusetts. The goats were born to two different mothers, but each goat was a clone, created in a similar way to Dolly the sheep. Mira 1 and her sisters, Mira 2 and Mira 3, were the first cloned goats. They were also the first transgenic goats (goats containing one or more genes from another animal). They contained a gene for producing a human protein called antithrombin III (ATIII) in their milk. ATIII plays an important role in blood clotting. Some people have a deficiency of this important protein, but it is also given to people in intensive care and is sometimes used during transplant operations. In the future, ATIII for medical use could be produced using transgenic goats like the three Miras.

Human clones?

Sheep, pigs, dogs, cats, and monkeys have been successfully cloned. So, is it possible to clone a human?

Many people are very strongly against the idea of human cloning, and in some countries laws have been passed to ban human cloning research. Even if people were not against the idea, most cloning scientists believe that it would be many years before it was safe to try to clone a human. In most animal cloning experiments, many embryos die before they can be born, and those that do survive have shortened lives. This would be unacceptable for a human clone.

Protesters outside a U.S. biotechnology company demonstrate against cloning research. Many people think that cloning humans is wrong under any circumstances.

Therapeutic cloning

It is possible to produce embryonic stem cells from eggs donated by women. The stem cells can then be grown in the laboratory. However, if the stem cells are to be useful for repairing damage, they would need to be genetically identical to the patient's cells. Cells with different genes would be rejected by the patient's immune system (the body's defense against disease).

One way to get around this problem might be to use nuclear transfer to clone embryonic stem cells. The nucleus from a cell in the patient's body could be added to a human egg cell with its nucleus removed. The egg cell with its new nucleus would then be grown until it was a ball of cells, when the stem cells could be harvested from it. These stem cells would have the patient's genes and could be used to treat the patient without fear of rejection.

CUTTING EDGE: STEM CELLS FROM SKIN

In November 2007, two separate groups of scientists managed to "reprogram" human skin cells to make them almost exactly like embryonic stem cells. Groups of scientists in Japan and the United States took skin cells from a human donor and used specially modified viruses to activate four genes in the cells. Once the genes had been turned on, the cells then behaved like embryonic stem cells.

They could be grown in the laboratory and converted into all kinds of different specialized cells, including muscle and nerve cells.

If this type of stem cell research proves successful, there may no longer be a need to use embryonic stem cells in research. This would remove the main ethical and religious objections against stem cell research.

Rabbits were first cloned in France in 2002. Like goats, sheep, and cows, they can be modified to produce drugs in their milk. →

Mouflons and gaurs

In 2001 a rare kind of ox, known as a gaur, was cloned. The embryo developed in a cow, rather than in a female gaur. The calf was named Noah, but unfortunately he died after only two days. Later that year, the first endangered animal was successfully cloned—a rare kind of sheep called a mouflon.

There is some doubt about whether cloning is a useful way of conserving endangered species. Creating clones of dead or very old animals could be a useful way of saving characteristics of rare animals that would be otherwise lost. However, cloning is not a substitute for other ways of protecting animals. Often, the main reason that a species is close to extinction (being wiped out) is the lack of suitable habitat, and cloning does not solve this problem.

Genetics in the Future

Genetic technology has already found many practical uses, ranging from producing life-saving medicines to helping make soft-centered chocolates. Today, scientists have only scratched the surface of what genetic technology could do. If some practical and safety problems can be solved, genetic technology has huge potential.

Many new uses for genetically modified plants and animals have already been tested experimentally. Some of these uses may become commonplace in the future. For example, it could become normal to farm animals or grow crops to produce drugs and medicines, in the same way that food, milk, or wool are produced today.

Spider silk (see page 31) could be just the first of many materials made by GM organisms. Polluting chemical factories and oil refineries could be replaced by farms producing materials such as plastics, high-tech fibers, and even fuels. Scientists have already genetically modified certain kinds of bacteria to produce gasoline. Other microbes called **algae** have been modified to produce hydrogen, which can also be used as a fuel and is much less polluting than gasoline. It is possible that hydrogen produced using genetic technology could help solve the world energy crisis.

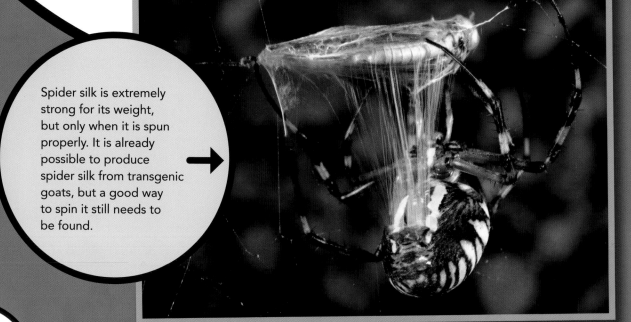

Spider silk is extremely strong for its weight, but only when it is spun properly. It is already possible to produce spider silk from transgenic goats, but a good way to spin it still needs to be found.

Stem cells

For years it has been suggested that stem cells could revolutionize medicine. Research is now at the stage where such a revolution seems a real possibility. Doctors may soon solve the problem of nuclear transfer in humans, so that embryonic stem cells can be made that match up with a patient's own genes. Illnesses such as heart problems and damage to the spine could be treated. Scientists may also be able to grow new organs for transplantation. Already, doctors have managed to grow artificial bladders that have been transplanted into patients with bladder problems. Another possibility is turning some of the patient's own cells into embryonic stem cells. (See the box on producing stem cells from skin on page 43.)

This micrograph shows a ball of embryonic stem cells. They can turn into any of the 200 different types of cell found in the human body.

CUTTING EDGE: RNA

Until recently, genetic research was focused on genes that produce proteins. However, more recently the focus has shifted to include other genes that produce RNA (another kind of nucleic acid—see page 18). Some kinds of RNA can act like enzymes and control chemical reactions in cells. Other kinds of RNA can shut down or start up genes. It is likely that RNA will become an important part of genetic technology in the future.

A gene discovered in fruit flies helps them live longer. It is called the Methuselah gene. →

Eternal youth?

One other area of genetic technology that holds great promise for the future is the genetics of aging. Although people in developed countries are living longer than at any time in the past, often they have a poor quality of life in old age. They suffer from all kinds of illnesses, which make it impossible to live a full life. Treating the illnesses of an increasing number of old people is also very expensive.

Scientists researching aging in animals have identified particular genes that are connected with aging. These genes seem to slow down aging and prolong life. For example, U.S. scientist Michael Rose has found a gene in fruit flies that helps them live longer. The gene seems to act by improving the body's defense and repair mechanisms. In the future, it may be possible to switch on genes in humans, or even to add genes from other animals that will help people age more slowly.

Opposition to genetic technology

Not everyone agrees that genetic technology is a good thing. Some people are concerned that GM plants and microbes might escape into the environment and become "super weeds" or "super bugs." Others are worried about the possibility of human cloning and of parents being able to choose "designer babies" with particular characteristics. As with any other powerful tool, genetic technology has the potential to be misused. However, many people conclude that the great benefits will outweigh the drawbacks, especially as understanding of genetic technology improves.

CUTTING EDGE: ARTIFICIAL LIFE?

Scientist and businessman Craig Venter has been at the cutting edge of genetics for many years. However, his ideas and research are often controversial. At present, Venter and his company, Synthetic Genomics, are taking on the ultimate challenge—they are working to design new forms of life. Venter's idea is to study the genetic makeup of many microbes to try to understand how they work. He will then use this information to make new microbes, designed to do things such as produce clean fuels and "scrub" excess carbon dioxide from the environment. Venter's vision is very grand and impressive. However, some scientists think that experiments with artificial life could cause more problems than they solve.

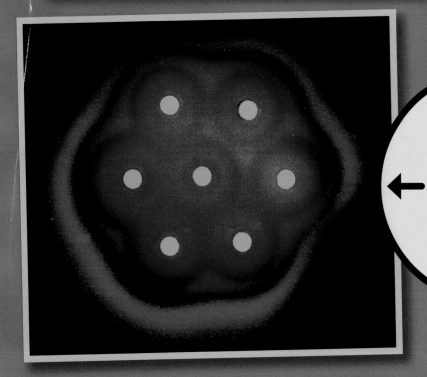

Bacteria have been designed to grow in a pattern in a shallow dish. Bacteria organized like this could be part of "living machines," designed to perform useful tasks.

Facts and Figures

Timeline of genetics history

1866 Gregor Mendel publishes a paper investigating the inheritance of characteristics in pea plants. This lays the foundations of the science of genetics.

1900 Carl Correns, Hugo de Vries, and Erich von Tschermak all independently rediscover Mendel's work, and modern genetics is born.

1902 U.S. biologist Walter Sutton shows that chromosomes exist in pairs. He suggests that hereditary factors must lie in chromosomes.

1909 Danish botanist Wilhelm Johannsen suggests the word "gene" for the hereditary factors described by Mendel. He also proposes two terms, "genotype" and "phenotype," for a person's genetic makeup and outward appearance.

1927 Hermann J. Muller uses X-rays to cause artificial gene mutations in fruit flies.

1951 Rosalind Franklin obtains sharp X-ray diffraction photographs of DNA.

1952 Martha Chase and Alfred Hershey carry out experiments that prove that DNA is the substance that transmits inherited characteristics from one generation to the next.

1953 British scientist Francis Crick and U.S. scientist James Watson discover the double-helix structure of DNA, using X-ray diffraction pictures from Rosalind Franklin and Maurice Wilkins.

1955 Indonesian biologist Joe Hin Tjio confirms that humans have 46 chromosomes (23 pairs). For 30 years, the number was believed to be 48.

1958 Matthew Meselson and Frank Stahl perform an experiment to prove that DNA replicates by building a new strand on each of the two original strands.

1958 Arthur Kornberg purifies DNA polymerase from the *E. coli* bacterium. It is the first enzyme in the genetic engineer's "toolbox."

1961 Sydney Brenner, François Jacob, and Matthew Meselson show that messenger RNA (mRNA) is the molecule that carries the genetic information from DNA in the nucleus out into the cell, where it is used to make proteins.

1972 Paul Berg and Herb Boyer produce the first recombinant DNA molecules (DNA with foreign genes inserted into them).

1977 Fred Sanger develops a method for sequencing DNA. In 1979 he discovers the sequence of a virus, phage-174. He does this entirely by hand.

1982 Drug company Eli Lilly markets the first genetically engineered medicine—human insulin made by recombinant bacteria.

1983 Scientists create genetically modified tobacco plants that are resistant to an antibiotic. The first transgenic plant (tobacco) is made.

1985 Kary B. Mullis invents the polymerase chain reaction (PCR), a way of "amplifying" small samples of DNA millions of times.

1989 Alec Jeffreys coins the term "DNA fingerprinting." He is the first to use the technique in paternity, immigration, and murder cases.

1990 James Watson and many others launch the Human Genome Project, a plan to map the entire sequence of human DNA.

1991 The first gene therapy trials are carried out on humans.

1994 The first GM plants (tomatoes) are sold in stores. The tomatoes contain a gene to help them stay firm.

1995 Craig Venter's company, Celera, discovers the sequence of the DNA of the bacterium *Hemophilus influenzae*.
A group of scientists sequences the complete genome of baker's yeast, which has more than 12 million base pairs.
GM soybean that is modified to be tolerant of herbicides is first sold in the United States.

1999 Jesse Gelsinger dies in a gene therapy trial.

2003 The Human Genome Project publishes the full human genetic sequence in the journal *Nature*, more than two years ahead of schedule.

2007 Two teams led by Shinya Yamanaka of the University of Kyoto, Japan, and James Thomson of the University of Wisconsin report the creation of "embryonic" stem cells from human skin cells.

Timeline of cloning history

1901 Hans Spemann splits a newt embryo that has just two cells, and grows two complete newt larvae.

1928 Hans Spemann carries out primitive nuclear transfer experiments involving salamander embryos. He suggests that the next step for research should be to clone organisms by extracting the nucleus of an adult cell and putting it into an egg with the nucleus removed.

1952 The first animal cloning begins. Robert Briggs and Thomas J. King clone northern leopard frogs.

1962 Biologist John Gurdon clones South African frogs using nuclei from adult cells.

1979 German biologist Karl Illmensee claims to have cloned three mice. His claim is later doubted, as he works in isolation and no one is able to repeat his experiment.

1984 Danish scientist Steen Willadsen makes a genetic copy of a lamb from early sheep embryo cells, a process now called "twinning."

1985 Steen Willadsen uses his cloning technique to duplicate prize cattle embryos.
Ralph Brinster creates the first transgenic animals—pigs that produce human growth hormone.

1993 Human embryos are first cloned using "twinning". (See above.)

1996 On July 6, Dolly the sheep, the first mammal ever to be cloned from adult cells, is born.

1998 Japanese scientists Ryuzo Yanagimachi and Teruhiko Wakayama in Hawaii clone 50 mice from an adult cell. Some of the mice are clones of clones. Japanese scientists make eight clones of a cow. This is only the third mammal species to be cloned.

2000 The group that created Dolly the sheep announces the first cloned pigs. Scientists hope that pigs could be genetically engineered for use in human organ transplantation.

2001 A cat, named C.C. for "carbon copy," is cloned by a company that wants to go into business reproducing pets.
A rare kind of ox, a gaur, is cloned. The surrogate mother is a cow. The gaur dies after only two days.
The first successful cloning of an endangered animal, a rare kind of sheep called a mouflon, occurs.

2005 South Korean scientists clone a dog named Snuppy. Dogs are particularly difficult to clone.

2007 A team of scientists in Oregon, led by Shoukhrat Mitalipov, produce 20 cloned macaque monkeys.

Find Out More

Further reading

Fridell, Ron. *Cool Science: Genetic Engineering*. Minneapolis: Lerner, 2006.

Morgan, Sally. *Chain Reactions: From Mendel's Peas to Genetic Fingerprinting: Discovering Inheritance*. Chicago: Heinemann Library, 2007.

Sneddon, Robert. *Cells and Life: Cell Division and Genetics*. Chicago: Heinemann Library, 2008.

Sneddon, Robert. *Cells and Life: DNA and Genetic Engineering*. Chicago: Heinemann Library, 2008.

Watson, James, D. *DNA: The Secret of Life*. New York: Knopf, 2003.

Watson, James, D. *The Double Helix: Personal Account of the Discovery of the Structure of DNA*. New York: Touchstone, 2001.

Yannuzzi, Della. *Great Life Stories: Gregor Mendel: Genetics Pioneer*. New York: Franklin Watts, 2004.

Websites

www.pbs.org/wgbh/aso/tryit/dna
Visit the DNA Workshop. Learn about DNA structure, replication, coiling, and protein synthesis through interactive activities.

http://biologica.concord.org/webtest1/web_labs_genophenotype.htm
Explore BioLogica's Genotype to Phenotype. Learn how genes can affect final appearance by studying unusual animals—dragons!

http://learn.genetics.utah.edu/units/basics/builddna
Think you know the structure of DNA? Check by building a DNA molecule for yourself.

http://learn.genetics.utah.edu/units/cloning/clickandclone
Click and clone. Use the techniques that produced the clone Dolly the sheep to clone Mimi the mouse.

http://learn.genetics.utah.edu/units/stemcells/index.cfm
This website is a good place to learn more about stem cells and stem cell research.

www.biotechnologyonline.gov.au/biotec/cloneanimal.cfm
Learn about the basic techniques of cloning, then try cloning your pet dog or bringing the extinct Tasmanian tiger back to life.

www.usoe.k12.ut.us/CURR/Science/sciber00/7th/genetics/sciber/punnett.htm
Learn how to use Punnett Squares to understand what happens when different alleles mix.

www.pbs.org/wgbh/nova/sheppard/analyze.html
Create a DNA Fingerprint and help to convict a killer.

Other topics to research

- Investigate the debate around the ethics of genetic research. Find two websites, one that is for genetic research and one that is against it, and summarize their views. Also, find out what the laws about genetic research are in your country. Would human cloning be allowed, if it were possible?

- Find out more about the ways that genetic technology affects your life. Look at the uses of enzymes made by genetic engineering. Find out more about the range of medicines produced using genetic technology. What are "monoclonal antibodies," and why are they important?

Glossary

alga (plural: algae)   plant-like creature, usually microscopic, that lives in water

allele   different "version" of the same gene

antibiotic   drug that kills bacteria or stops them from growing

atom   very tiny particle that everything is made from

bacterium (plural: bacteria)   simple living thing made of just a single cell. Unlike most other organisms, bacteria do not have a definite nucleus or organelles.

cell   basic unit of living things

cell wall   tough outer layer found in plant cells

chromosome   long, thin structure seen in the nucleus of a living cell. Each chromosome contains a single, tightly coiled DNA molecule.

clone   genetically identical copy of a living thing

codon   group of three DNA or RNA bases; the basic unit of the genetic code

DNA   (deoxyribonucleic acid) genetic material of living things

dominant   allele that is always outwardly expressed when present in the genetic makeup of living things

embryo   new life in the early stages of development, before it is born or has hatched from an egg

enzyme   specialized protein that speeds up chemical reactions in living cells

fermentation   when bacteria or other microbes are grown in conditions where there is food, warmth, and little or no oxygen

fertilization   when male and female sex cells (gametes) join to form a zygote

gene   unit of heredity—a section of DNA that codes for the production of a protein or a small group of proteins

gene therapy   curing a genetic disease by reinforcing the patient's damaged genes with undamaged versions

genome   all the genetic material of a living thing

grafting (in plants)   way of growing a new plant by joining the roots of one plant to a piece of another plant, which then grows a stem, leaves, and flowers

hemoglobin   red pigment responsible for carrying oxygen in the blood

heterozygous   having two different alleles of a particular gene

homozygous   having two copies of the same allele of a particular gene

ligament   tough fiber that connects bones together

livestock   farm animals such as cattle, sheep, pigs, and goats

malaria   infectious disease, usually found in warm countries, transmitted by a type of mosquito

microbe   microscopic living thing

micrograph   photograph taken through a microscope showing the magnified image

molecule   group of atoms joined together through chemical bonds

nuclear transfer   process in which the nucleus from a cell is inserted into an egg cell that has had its nucleus removed

nucleotide base (or simply "base")   small molecule that is the building block for DNA. There are four different bases in DNA.

organelle   small structure within a living cell

plasmid   circular piece of DNA found in bacteria

recessive   allele that is outwardly expressed only when two of the same allele are present in an organism's genetic makeup

ribosome   tiny structure in a cell that is the site of protein synthesis

sequencing   figuring out the order of base pairs along the length of a DNA molecule

species   group of similar living things that can breed together

staple food   main food of people in a particular area

stem cell   cell that retains the ability to divide and differentiate (change) into several different kinds of specialist cell

tendon   stringy, slightly stretchy fiber that connects muscles to bones or to other muscles

tissue   group of cells in the body that do a similar job

transgenic   organism that has genetic material from another living thing incorporated into its DNA

variation   genetic differences among individuals of the same species

virus   very tiny, simple particle that can use the machinery of living cells to reproduce itself

yeast   microscopic living thing that is a type of fungus

zygote   fertilized egg cell—the first cell of a new living thing

Index

aging process **46**
alleles **11, 12, 13, 14, 15, 21**
  dominant **12, 14**
  recessive **12**
amino acids **17, 18, 20**
animals
  GM animals **4, 29, 31, 37, 44**
  transgenic animals **31, 40, 41**
antibiotics **30**
antibodies **20**
artificial life **4, 47**
atoms **16**

bacteria **25, 47**
  GM bacteria **26, 27, 28, 34, 35, 44**
  medicines from **27**
blood clotting **27, 31**
blood types **14**

cells **8, 37**
  division **8, 22**
  sex cells **10–11**
  skin cells **43**
  stem cells **41, 42, 43, 45**
chromosomes **23, 48**
cloning **4, 38–43, 50–51**
  animals **38, 39–40, 41, 42, 43, 50–51**
  artificial **38–41**
  humans **42**
  natural **38**
  plants **33, 39, 40**
  therapeutic **42**
codons **17**
criminology **19**
cystic fibrosis **28, 30**

DNA (deoxyribonucleic acid) **8, 9, 10, 11, 16–25, 30, 48**
  bacterial DNA **26, 27**
  DNA array **29**
  DNA fingerprinting **19, 24, 25, 49**
  double-spiral structure **16, 48**
  mutations **17**

sequencing **23, 49**
  sub-units **16, 17, 18**
DNA ligase **22, 25, 26**
DNA polymerases **22, 24, 48**
  polymerase chain reaction (PCR) **24, 49**
Dolly the sheep **40**

earlobes **12**
enzymes **20, 22, 25, 37, 48**
  restriction enzymes **25, 26**
eye color **11**

fermentation **27**
"Frankenstein science" **5, 9**

gene guns **35**
gene therapy **30–31**
genes **10–15, 20**
  alleles **11, 12, 13, 14, 15, 21**
genetic code **17**
genetic disorders **28–29, 30**
genetic engineering **4, 5, 9**
genetic technology
  gene therapy **30–31**
  GM animals **4, 29, 31, 37, 44**
  GM medicine **26–29**
  GM plants **4, 5, 31, 32–36, 44, 49**
  opposition to **37, 47**
genetics **6–9**

hemoglobin **21**
hemophilia **27, 31**
heredity **6, 7, 8**
heterozygous **21**
homozygous **21**
human embryos **4, 29, 42**
Human Genome Project **23, 49**

insulin **28**

liposomes **30**

Mendel, Gregor **7, 12, 13, 48**
Methuselah gene **46**

mice
  glow-in-the-dark mice **4**
  knock-out mice **29**
microbes **4, 27, 47**
molecules **16, 20, 22, 48**

nuclear transfer **38, 45**
nucleotide bases **16, 17, 22, 23**

organ transplants **45**

plants
  breeding **6, 7, 12–13, 14, 33**
  cloning **33, 39, 40**
  GM plants **4, 5, 31, 32–36, 44, 49**
plasmids **26, 27**
proteins **17, 18, 20, 21**
  fibrous **20**
  globular **20**
Punnett squares **15**

restriction enzymes **25, 26**
RNA **18, 45**

selective breeding **6, 33**
sexual reproduction **10**
sickle cell disease **21, 28**
spider silk **31, 44**
stem cells **41, 42, 43, 45**
  embryonic **42, 43, 45**

transgenic animals **31, 40, 41**
twins, identical **38**

viruses **17, 30, 31**

zygotes **8**